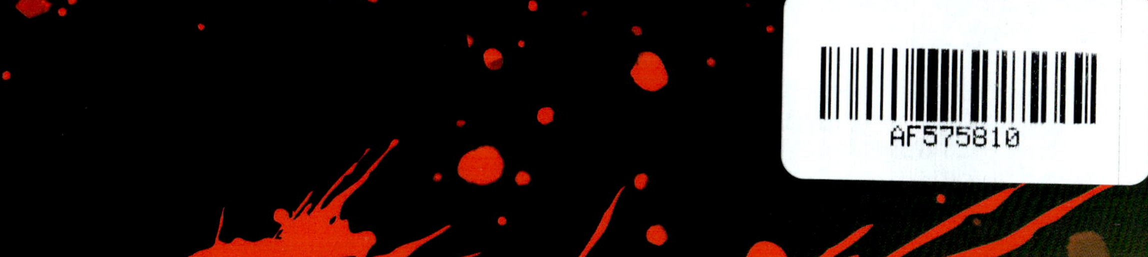

DEADLIEST ANIMALS

KILLER BEES

A Crabtree Branches Book

Amy Culliford

School-to-Home Support for Caregivers and Teachers

This high-interest book is designed to motivate striving students with engaging topics while building fluency, vocabulary, and an interest in reading. Here are a few questions and activities to help the reader build upon his or her comprehension skills.

Before Reading:

- *What do I think this book is about?*
- *What do I know about this topic?*
- *What do I want to learn about this topic?*
- *Why am I reading this book?*

During Reading:

- *I wonder why…*
- *I'm curious to know…*
- *How is this like something I already know?*
- *What have I learned so far?*

After Reading:

- *What was the author trying to teach me?*
- *What are some details?*
- *How did the photographs and captions help me understand more?*
- *Read the book again and look for the vocabulary words.*
- *What questions do I still have?*

Extension Activities:

- *What was your favorite part of the book? Write a paragraph on it.*
- *Draw a picture of your favorite thing you learned from the book.*

TABLE OF CONTENTS

KILLER BEE

There are nearly 20,000 **species** of bees worldwide. The most familiar is the honey bee. Honey bees produce a thick, golden, sweet treat people have enjoyed for thousands of years.

In the 1950s, scientists began **cross-breeding** honey bees. Their goal was to breed a more efficient honey maker. They met their goal, but also created a dangerous **hybrid**. The killer bee was born!

Why are they called killer bees? The hybrid bees are more **defensive** than other honey bees. If disturbed, the bees aggressively **swarm**, chase down, and sometimes kill the intruder.

The killer hybrid is also called the Africanized honey bee. It was created when African honey bees were brought to South America and bred with European honey bees. Some of the Africanized bees escaped and spread north. Killer bees are now found as far north as Texas, California, and Florida.

An Africanized honey bee (left) and a European honey bee. They look similar, but honey bees can vary in color. Often, the only way to tell Africanized honey bees from other bees is in a lab.

KILLER LOOKS?

Killer bees look like other honey bees. Their hairy bodies are golden yellow with brown bands. And, like all insects, killer bee bodies have three segments.

KILLER BEES ARE SMALLER THAN MOST OTHER HONEY BEES. THEREFORE, KILLER BEES CARRY LESS VENOM.

The first segment is the head, where there is a set of antennae. The second segment is the thorax, where two sets of wings and three sets of legs are attached. The third segment is the abdomen, or tail end, where the stinger is located.

THE COLONY

Killer bees live in **colonies.** A colony has one queen bee, worker bees, and drones. The queen's colony will serve and protect her with their lives. A colony builds its own home, or nest. Hollow trees or rocky nooks make good nest sites. But, beware! Killer bees can also build nests in old tires, behind house siding, or in junked cars.

killer bee nest

Nests close to where people are active can be very dangerous. Killer bees do not like to be disturbed.

HOMES BUILT BY BEES ARE CALLED NESTS. HUMAN-MADE HOMES FOR BEES ARE HIVES.

THE DRONES

Drones are male bees. They are easily spotted because they have bigger eyes than their nest mates. Drones do not have stingers. Their only job is to mate with the queen.

The lifespan of a killer bee is not long. Drones live one or two months. Worker bees live only about one month, while the queen can live about three years.

THE WORKERS

All worker bees are female. Workers take on the hardest jobs in the colony. They spend their short lives in devoted service to the queen. Worker bees bring in water and search for food.

They feed the queen and drones, guard the colony, and even take out the trash by removing waste from the nest. Workers also **regulate** the nest's temperature. They flap their wings to move the air when it's warm, and use body heat to warm the nest when it's cold.

THE QUEEN

The queen is the most important bee in the colony. She is the only bee that can reproduce. She can lay 1,500 eggs a day for about three years. The eggs hatch into bee **larvae** after three days.

Beekeepers usually mark the queen with a dab of paint.

bee eggs

When the queen grows old, or dies, worker bees select larvae to be potential new queens. Workers feed larvae royal jelly—food made by the worker bees. The jelly helps the larvae develop into queens. From these queens, a dominant bee will rise. She will sting and kill any competitors. There can be only one queen bee!

HONEY

Killer bees are great honey producers. Some beekeepers even prefer them. Why do bees make honey? Honey is bee food. Bees gather **nectar** from flowers. Enzymes in their stomach break down the nectar and remove water. The bees store the processed liquid, now honey, in the cells of the honeycomb. Stored honey helps bees live through the winter. When beekeepers harvest, they leave enough honey for the bees to eat.

THE HONEYCOMB IS THE WAX STRUCTURE MADE BY BEES. IT STORES HONEY AND PROVIDES A SAFE HOME FOR LARVAE TO DEVELOP.

KILLER WEAPONS

Weapon 1—The Stinger

stinger

The stinger is a pointed tube with barbs. The barbs grab onto skin and get stuck. The bee cannot withdraw the stinger, so it leaves it behind along with a venom sac and other parts of the abdomen. Because of this, a killer bee can only sting once, and then dies.

A queen will sting other potential queens in her nest. Her stinger is slightly smooth, but still barbed. She doesn't lose her stinger when she attacks, so she is capable of stinging multiple times.

WEAPON 2—VENOM

Killer bee venom is a colorless liquid. Bees release venom through their stinger. Venom can cause serious pain and even life-threatening reactions in some people.

WEAPON 3—NUMBERS

Killer bees are dangerous because they attack in great numbers.

Killer bee stings release special chemicals along with their venom. The chemicals act as a signal for other bees to sting.

KILLER BEE STING

A killer bee sting can cause local pain, itching, and swelling. For some people, severe allergic reactions can occur. Severe reactions can cause difficulty breathing and heart, liver, or brain damage. Some people can even go into shock and die.

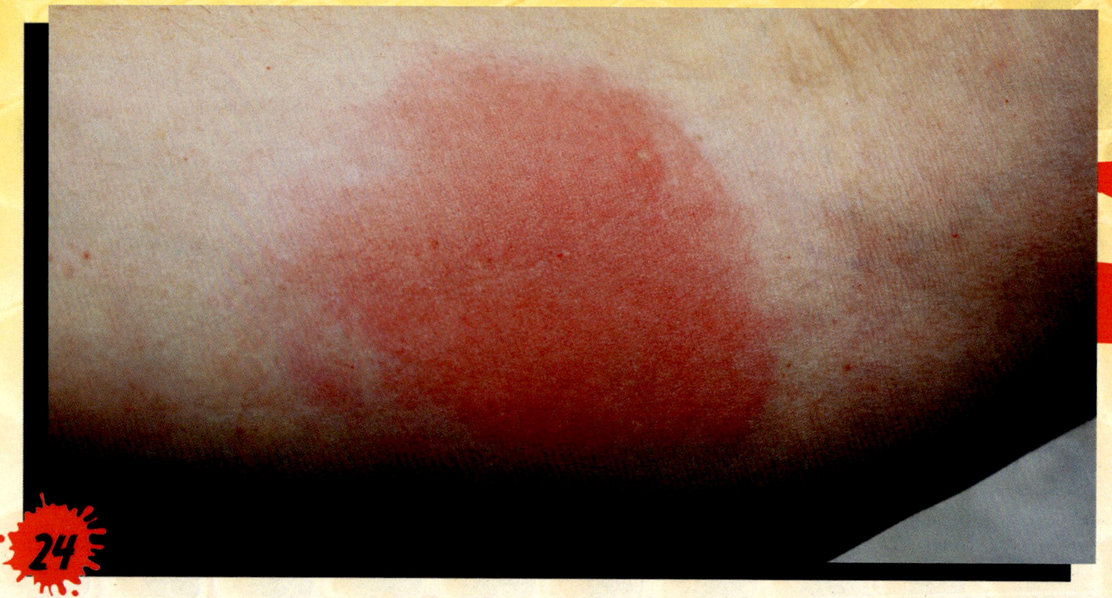

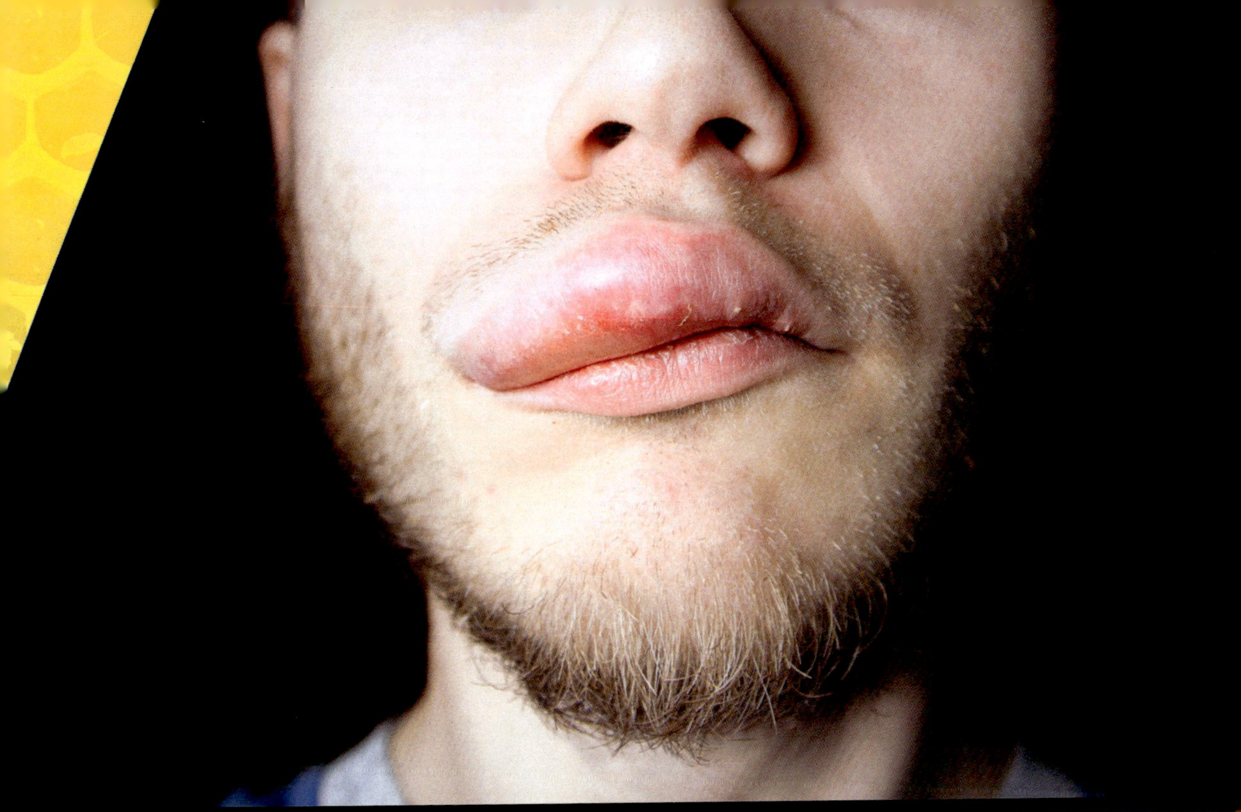

People who have severe reactions to a bee sting benefit from a shot of the hormone epinephrine, also known as adrenaline. The hormone can be carried in a self-injectable pen for emergency use.

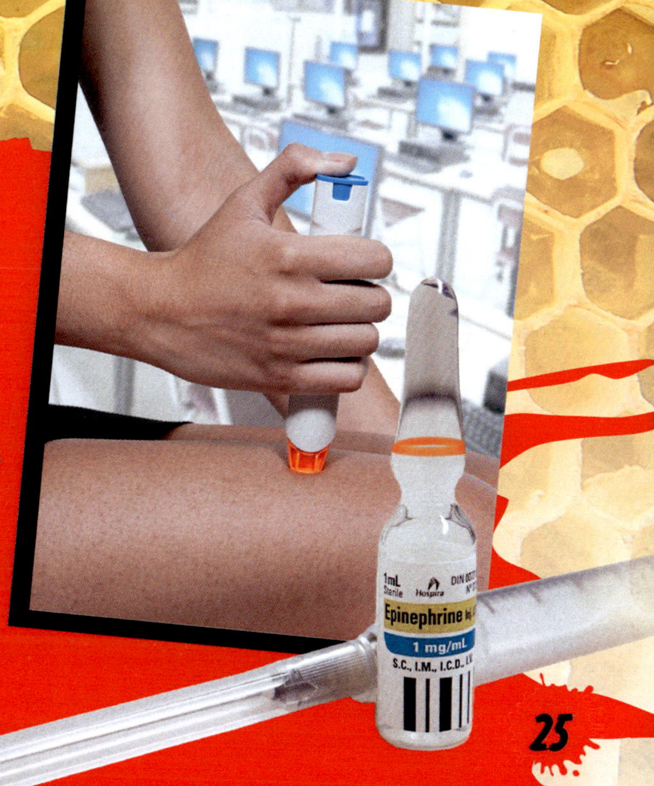

KILLER BEES ATTACK!

Killer bees attack anything that threatens their colony. Once a threat is identified, worker bees go into a frenzy. They attack in force, and will chase their target a quarter mile (0.4 km) or more. One attack occurred in Texas in 2013. A man

accidentally disturbed a killer bee nest while driving his tractor. The colony swarmed, stinging the man over a thousand times. Family and neighbors were also stung as they tried to help. Unfortunately, the man died from the massive attack.

Killer bees kill other insects, small animals, and even large animals like dogs and horses. Killer bees have even killed hundreds or possibly thousands of people worldwide. When killer bees feel threatened, they attack in great numbers, overwhelming their victim. That's why killer bees are one of the world's most deadliest animals.

Bee exterminators wear protective clothing when removing bee nests.

Dangerous nests should be removed from populated areas.

GLOSSARY

apiphobia (eh-pif-OH-bee-uh): Intense fear of bees

beekeepers (BEE-keep-erz): People who keep bees, honey farmers

colonies (KOL-uh-NEEZ): Large groups of bees or other insects

cross-breeding (KRAWSS-BREE-ding): Mating two different kinds of animals or plants to produce a new kind

defensive (di-FEN-siv): Feeling and acting as if being attacked

epinephrine (ep-uh-NEF-rin): A hormone released in the body in response to stress that increases heart rate, muscle strength, and blood sugar

hybrid (HYE-brid): Bred from two different species

larvae (LAR-vee): Insects that look like worms in a stage of development after egg and before adult

nectar (NEK-tur): A sweet liquid bees collect from flowers and turn into honey

regulate (REG-yuh-late): Control or manage

species (SPEE-sheez): A certain type of animal or plant

swarm (SWORM): A thick mass of insects that moves in large numbers

venom (VEN-uhm): Harmful fluid delivered by injection

Index

Websites to Visit

www.si.edu/spotlight/buginfo/killbee

https://kids.kiddle.co/Africanized_honeybee

www.pestworldforkids.org/pest-guide/bees/

About the Author

Amy Culliford

Amy Culliford has a Bachelor of Fine Arts. She has worked as a classroom drama teacher and led after-school drama programs. She avoids bees of any kind.

The author would like to thank David and Patricia Armentrout for their research and help on this project.

Produced by: Blue Door Education for Crabtree Publishing
Written by: Amy Culliford
Designed by: Jennifer Dydyk
Edited by: Tracy Nelson Maurer
Proofreader: Crystal Sikkens

Photographs: Cover photo © Felipe Duran/Shutterstock.com, graphic splat on cover and throughout © Andrii Symonenko /Shutterstock.com, page background photo of honeycomb throughout © Kostiantyn Kravchenko, page 4 honeycomb © Dionisvera/Shutterstock.com, page 5 bees © StudioSmart/Shutterstock.com, page 6 © Pamela Au,Shutterstock.com page 7 courtesy of the USDA, page 8 © mikeledray/Shutterstock.com, page 9 © Jaco Eksteen/Shutterstock.com, page 11 (top) © © Ktr101 https://creativecommons.org/licenses/by-sa/4.0/deed.en, (bottom) © CatherineLProd/Shutterstock.com, page 12 and 13 bees © Kuttelvaserova Stuchelova/Shutterstock.com, page 14 (top) and 15 (top) © Kuttelvaserova Stuchelova/Shutterstock.com, page 14 (bottom) © Simon_g/Shutterstock.com, page 15(bottom) © Ruth Swan/Shutterstock.com, page 16 both photos © Kuttelvaserova Stuchelova/Shutterstock.com, page 17 © bee eggs © Jay Ondreicka/Shutterstock.com, larvae © Kuttelvaserova Stuchelova/Shutterstock.com, life cycle illustration © EreborMountain/Shutterstock.com, page 19 (top) © New Africa/Shutterstock.com (bottom) © Ian Scammell/Shutterstock.com, page 20 © Katrina Brown/Shutterstock.com, page 21 © Bachkova Natalia/Shutterstock.com, page 23 large photo © Mirko Graul/Shutterstock.com, small bees © Katrina Brown/Shutterstock.com, page 24 © Siegi/Shutterstock.com, page 25 (top) © Velimir Zeland/Shutterstock.com, girl with epi-pen © Rob Byron/Shutterstock.com, epinephrin vial © Editorial credit: Sheriffhay / Shutterstock.com, page 26 © Pamela Au/Shutterstock.com, page 27 © DenysHolovatiuk/Shutterstock.com, page 28 and 29 (top) © Editorial credit: mikeledray / Shutterstock.com, page 29 (bottom) © AlexGD93/Shutterstock.com

Library and Archives Canada Cataloguing in Publication

Title: Killer bees / Amy Culliford.
Names: Culliford, Amy, 1992- author.
Description: Series statement: Deadliest animals | "A Crabtree branches book". | Includes index.
Identifiers: Canadiana (print) 20210218894 | Canadiana (ebook) 20210218908 | ISBN 9781427154149 (hardcover) | ISBN 9781427154200 (softcover) | ISBN 9781427154262 (HTML) | ISBN 9781427154323 (EPUB) | ISBN 9781427154385 (read-along ebook)
Subjects: LCSH: Africanized honeybee—Juvenile literature.
Classification: LCC QL568.A6 C85 2022 | DDC j595.79/9—dc23

Library of Congress Cataloging-in-Publication Data

Names: Culliford, Amy, 1992- author.
Title: Killer bees / Amy Culliford.
Description: New York : Crabtree Publishing, 2022. | Series: Deadliest animals - a Crabtree branches book | Includes index.
Identifiers: LCCN 2021022052 (print) | LCCN 2021022053 (ebook) | ISBN 9781427154149 (hardcover) | ISBN 9781427154200 (paperback) | ISBN 9781427154262 (ebook) | ISBN 9781427154323 (epub) | ISBN 9781427154385
Subjects: LCSH: Africanized honeybee--Juvenile literature. | Dangerous animals--Juvenile literature.
Classification: LCC QL568.A6 C78 2022 (print) | LCC QL568.A6 (ebook) | DDC 595.79/9--dc23
LC record available at https://lccn.loc.gov/2021022052
LC ebook record available at https://lccn.loc.gov/2021022053

Crabtree Publishing Company
www.crabtreebooks.com 1-800-387-7650

Published in the United States
Crabtree Publishing
347 Fifth Avenue, Suite 1402-145
New York, NY, 10016

Published in Canada
Crabtree Publishing
616 Welland Ave.
St. Catharines, ON, L2M 5V6

Printed in Canada/112022/CPC22021123